AF853408

Mudflat Mania!

Copyright © 2009 Rowley Bridge Press, Hastings-on-Hudson, NY and
the Cape Cod Museum of Natural History, Brewster, MA

For information about permission to reproduce selections
from this book, email: info@rowleybridge.com

Cape Cod Museum of Natural History and the heron logo are
trademarks of the Cape Cod Museum of Natural History, Brewster, MA

Published by Rowley Bridge Press, Hastings-on-Hudson, NY, 2009
Second printing, May 2010

Library of Congress Control Number: 2009921203

ISBN: 978-0-615-23223-2

Printed in the United States of America

Printed on SAPPI Opus paper certified
by the Forest Stewardship Council

Cape Cod Museum of Natural History

Mudflat Mania!

Exploring the ocean when the tide pulls out!

Written by Irene Ledwith

Introduction by Betsy and Bud Ferris

Rowley Bridge Press

An indoor and outdoor museum! The museum conducts guided field walks on three nature trails. One walk is the always popular *Mudflat Mania!*

The Cape Cod Museum of Natural History

in Brewster, Massachusetts, was founded in 1954. Its mission is to encourage and advance understanding and stewardship of the natural environment through discovery and learning. The museum houses two floors of exhibits about the flora and fauna of Cape Cod, including exhibits on whales, indigenous birds, and coastal change. Aquaria display different species of crustaceans, mollusks, fish, frogs, and turtles. The museum presents its own collections as well as relevant traveling and loan exhibits. Also, the museum acts as a steward of 300-plus acres of museum-owned land in Stoney Brook Valley and Brewster Conservation land adjacent to the museum.

All year round, the museum offers lectures and workshops for children and adults. In addition, the museum conducts guided field walks on three nature trails, which meander through woodland, salt marsh, and the shore of Cape Cod Bay. One of these walks is the always popular *Mudflat Mania!*

Mudflat Mania! was created by museum volunteer and teacher Claudia Kren. Claudia, a Ph.D. in the History of Science, was an avid naturalist. She was always ready to share her vast knowledge of invertebrates with museum guests eager to explore the tidal creeks. Thanks to Claudia's efforts, museum visitors of all ages can enjoy the fun and frolic of beachcombing while gaining an understanding and awareness of our changing natural environment. In addition to her work with *Mudflat Mania!*, Claudia contributed immensely to the museum's Marine Room and Field Guide Program, and developed the popular seminar series Dialogues on Darwin. Claudia died in January 2008. This book is dedicated to her.

Cape Cod Museum of Natural History
869 Main Street/Route 6A, Brewster, MA, 02631
508-896-3867
www.ccmnh.org

Large footsteps to follow! One museum member said of the guide work of Claudia Kren (above): "If you have ever taken a field trip with Claudia, you will never see the sand flats in the same way."

Science or play? Armed with shovels and pails, visitors to the Cape Cod Museum of Natural History are escorted by volunteer guides onto the milewide Brewster mudflats.

Contents

Introduction

Mudflat Mania is a popular field walk at the Cape Cod Museum of Natural History in Brewster, Massachusetts. Participants of all ages, armed with pails and shovels, walk with volunteer naturalists to the mudflats on Cape Cod Bay. Here, the eager seekers fondly known to the museum staff as Mudflat Maniacs look for and learn about marine invertebrate animals.

On these mudflats, the view to the edge of the sea at dead low tide extends across sandbars and tidal pools for one mile from the wrack line left by high tide only six hours before. This stretch of shoreline between Quivett Creek and Paine's Creek is Town of Brewster Conservation Land, set aside to protect a vast salt marsh and barrier beach, where an amazing diversity of life has adapted to diurnal and seasonal changes between a land world and an ocean world.

This guide provides brief descriptions and pictures

A familiar pose! With their heads and shoulders bent toward the earth, Mudflat Maniacs are easily identified. They are commonly huddled in groups.

of some of the animals commonly found here. We hope it will encourage people of all ages to experience Mudflat Mania and to discover and identify these creatures.

Mudflat Maniacs do not remove any natural thing from the environment. After sharing bucketsful of discovery with others, they return all creatures to their own habitat. Perhaps this book may serve as a better memento than a periwinkle's shell, which may be recycled as a new house for a hermit crab!

The animals in this guide represent only a fraction of those we have encountered over several years of field walks. Once, we discovered a nudibranch, but we haven't seen one since. Animals appear, or don't appear, according to their own ways. That makes each walk out on the mudflats an adventure, where the wonder of discovery waits for everyone who comes along.

— Betsy and Bud Ferris
Volunteer Naturalists
Cape Cod Museum of Natural History

The Tide

While enjoying a day at the ocean, we witness the steady change in the ocean tide. As high tide approaches, beachgoers move their chairs further back on the shore to avoid the encroaching water. Sandcastles built 20 feet from the water's edge, to the delight of their architects, are destroyed by ocean waves.

This tidal change is the result of the changing positions and gravitational attraction of the moon, sun and earth. As the moon orbits the earth, the moon's strong gravitational pull causes the earth's ocean water to bulge toward it. At the location of this bulge, it is high tide. At the same time, on the opposite side of the world, a second high tide occurs where the water is far less attracted. Low tide happens in between the two bulges. Because the moon takes 24 hours and 50 minutes to orbit the earth, high tide occurs 50 minutes or so later each day.

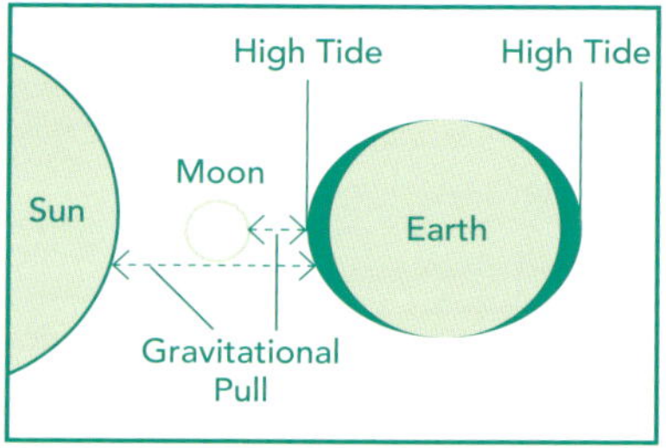

The Highest Tide. When the sun and moon are aligned for a new moon (as shown above) and a full moon, their combined pull creates the strongest tide, a "Spring tide." Its name refers to the *leap* in water.

Now you see it. . . . At low tide, the Brewster mudflats stretch a mile to Cape Cod Bay. In the photo at right the wrack line, the line of seaweed and other debris left by high tide, shows the ocean's reach just six hours prior.

Green Crab

A stowaway? The European Green Crab first entered the U.S. in the mid 1800s, most likely by sailing ship from Europe to the Cape Cod region. In the early 1900s they spread northward, arriving in Maine in the 1950s.

Also called European Green Crab

Name: *Carcinus maenas*

Size: Up to 3 inches in shell width

Attributes: The Green Crab is a voracious predator that feeds on mollusks, worms, and other crustaceans. It is distinguished by the three spinelike protrusions between the eyes, and five on each side of the carapace, the dorsal shell. It can vary in color from reddish to a mottled green. Its last pair of legs are somewhat flattened but not paddle-shaped. It is a large, active crab that matures at two to three years of age, lives up to five, and breeds up to three times a year.

Range: New Jersey to Nova Scotia. Since 1989 on the West Coast, now ranging there from California to British Columbia. It has been introduced to the coasts of South Africa and Australia as well.

Where am I? Look for these crabs scurrying about in tidal creeks. They are common at low tide. Their molts are often found washed ashore.

MANIAC ALERT

Approach a Calico Crab with care. It is an aggressive crab with quick reflexes. As seen in the photo above, it walks about on its rear legs with its long, sharp, serrated pinchers positioned up and ready to grasp an unsuspecting victim.

Calico Crab

Also called Lady Crab or Sand Crab

Name: *Ovalipes ocellatus*

Size: Up to 3 inches in shell width

Attributes: The Calico Crab is pale beige in color, with purple-red speckled clusters. It has five spinelike protrusions on each side of the carapace, about equal in size, and three between its eye sockets. The last pair of legs are flat and paddle-shaped, which allows this crab to swim. It is an ill-tempered crab with strong pinchers. It likes to bury itself in sand with its eyestalks exposed looking for prey. It feeds on live animals as well as decaying organisms. To identify its gender, look at its underside. The male has a pointed, oblong shape on its underbelly, resembling the Washington Monument, while the female's underbelly has a dome shape like the U.S. Capitol building. We're not kidding!

Range: Cape Cod to Gulf of Mexico

Where am I? This crab likes the sandy bottoms of tidal creeks. Its molts are commonly found washed ashore.

Leftover lunch! The Sand Fiddler Crab eats mainly bacteria and algae found in the sand. Once the crab has removed its food, it deposits the remaining soil in small balls just outside its burrow.

Sand Fiddler Crab

Name: *Uca pugilator*

Size: Up to 1 inch in shell width

Attributes: The Sand Fiddler Crab carapace (the large dorsal shell) is somewhat squarish, pinkish-purple in color, with a bright patch of purple in the center. The legs range in color from orange to brown. Its eyestalks are relatively short. It is identified chiefly by the male, which has an unusually large single front claw (while the female has two similarly sized small claws.) In an unusual mating ritual, the male waves the large claw while making noise to attract the female. Despite the large claw, fiddler crabs are relatively easy to handle. Just scoop them up gently with your bare hands.

Range: Cape Cod to Texas

Where am I? Its burrows can be found in the sand, often in areas with sea grass. Look for finger-size holes surrounded by small balls of sand. A related species, the Mud Fiddler can be found further back from the shore in the mud. It is darker in color by comparison.

Here's looking at you! The Hermit Crab is an active crab that will tirelessly scour the bottom of mudflats. With alert eyestalks, it looks for food to eat and abandoned shells that might make a new home. The crab in the photo above has found a camera lens.

Hermit Crab

Name: *Pagurus* (multiple species)

Size: Up to 2 inches in length, depending on species

Attributes: Hermit crabs are recognized by the homes they adopt to protect their unarmored abdomen. The Long-Clawed Hermit Crab is one of the most common hermit crabs found in Atlantic waters. Its homes of choice on Cape Cod are periwinkle, mudsnail, and moonsnail shells. Its body is light pink in color, and it has a pair of very narrow claws on the first legs. Its body has adapted so well to living in snail shells that its abdomen is twisted. This allows it to fit around the spire in the center of snail shells. It is an active but timid crab. It will withdraw into its adopted home when disturbed, leaving its large claw at the opening as a barrier.

Range: Massachusetts Bay to Eastern Florida and the Gulf of Mexico

Where am I? If you see a snail moving faster than at a snail's pace, it's a good bet it's a Hermit Crab. At low tide these crabs are abundant in tidal pools.

Do Sand Shrimp eat sand?
Well, not quite. Their diet consists mainly of detritus, the decomposed plant and animal matter broken down by the action of water and deposited in the sand.

Sand Shrimp

Also called Seven Spine Bay Shrimp

Name: *Crangon septemspinosa*

Size: Up to 2 1/2 inches in body length

Attributes: The Sand Shrimp is light tan in color, almost transparent with brown patches. This species is identified by the distinct claws on its first pair of walking legs. It is a bottom dweller and will burrow for food. Its spawning season ranges from March to October. The Sand Shrimp is the most common shrimp in the New England area and a favorite food to many carnivores, including weakfish and bluefish. A related species found on Cape Cod is the Grass Shrimp.

Range: Eastern Florida to the Arctic, and also found in the North Pacific

Where am I? These transparent animals are found in tidal creeks and pools, preferring areas with small rocks and shell debris, as in the photo at right. If you can't find the two shrimp in the photo, look in the upper right corner for one, and just left of center for the other.

Uninvited guests! Barnacles will attach themselves to any solid object that is not moving too fast. The Ribbed Mussel, shown in the photo above, has become a host to three. Barnacles prefer to settle with other barnacles, and find each other by a chemical odor.

Barnacle

Name: *Balanus* (multiple species)

Size: Up to 1 inch in shell diameter

Attributes: Barnacles are the only stationary crustaceans. They attach themselves to rocks, logs, shellfish, whales, and to the distress of mariners, to boats too. They always congregate in groups. The home of the Acorn Barnacle and the Northern Atlantic Barnacle looks like a small white igloo. They create this shape by secreting calcium-containing plates, which encase the animal. The animal itself is shrimplike. To feed, the barnacle will extend its feathery limbs into the water and trap plankton. Barnacles secrete an incredibly strong glue, which attaches them to a solid object, so they are very hard to remove. Having both female and male sex organs, they are hermaphrodites but must be fertilized by a neighboring barnacle.

Range: Cape Cod to Florida. Also in Europe.

Where am I? Look for little white igloos on anything not moving. When you find one, you find many.

Medical Miracle! Each year over 300,000 horseshoe crabs are harvested for a unique compound found in their blood that is used to detect harmful bacteria in drugs and on medical equipment. They are bled for this compound and then returned to the sea. Efforts are being made to protect horseshoe crabs, for their sake and ours.

Horseshoe Crab

Name: *Limulus polyphemus*

Size: Up to 2 feet in length, including the tail

Attributes: Though named a crab, it in fact is not one. It is an arthropod related to spiders. It is referred to as a "living fossil" because its genus, *Limulus,* dates back 350 million years to the Triassic period, the first period of dinosaurs. The Horseshoe Crab is dark brown in color with some black tones. It has a long spinelike tail. Though its armor looks forbidding, it is harmless. In the spring, when the species mates, the male, smaller than the female, will clasp onto the female and follow her to the sand in the uppertidal zone (as shown in the photo at right.) There, the female will deposit her eggs and the male will fertilize them. For food, they search the sand for worms, mollusks, and detritus.

Range: Maine to the Yucatan Peninsula

Where am I? Horsehoe Crabs are most commonly seen between May and June at the full-moon tide, when they venture ashore to mate.

MANIAC ALERT

In the photo above, the four horseshoe-shaped gonads and the faint outline of the body are all that visibly identify this moon jelly washed up on shore. But beware, though the jellyfish may be dehydrating, its stinging cells can still sting!

Moon Jelly

Name: *Aurelia aurita*

Size: The umbrella grows up to 10 inches across. The tentacles are relatively short in length.

Attributes: The Moon Jelly is translucent, and its color is either white, light pink, or beige. It has four distinct horseshoe-shaped gonads. It moves by pulsations of the umbrella-shaped upper body. It has around 250 hairlike tentacles that hang from the edge of its saucer-shaped umbrella. The tentacles are mucous-covered and catch food, which is wiped off and into the mouth by four oral lobes that hang down under the umbrella. It is a carnivore that feeds on minute organisms. The Moon Jelly is only slightly venomous, but it may produce a slight burning sensation upon contact with skin. It is one of the most common jellyfish on Cape Cod.

Range: Atlantic Ocean, Pacific Ocean, Indian Ocean

Where am I? Moon Jellies are common in the late spring and summer months and most often found washed up on shore or on sandbars.

Lion's Mane

Also called Red Jelly

Name: *Cyanea capillata*

Size: On Cape Cod, grows up to 12 inches wide. However, the Lion's Mane is considered the world's largest jellyfish. Its size varies by habitat, with the largest of the species found in the deep waters of the Gulf of Maine, where it can grow up to 8 feet wide!

Attributes: Its color varies by habitat also. Around Cape Cod, its color ranges from yellow to an orangy brown. Its bell is divided into eight lobes, and its long and sticky tentacles grow in eight clusters. Its name is derived from these long, colorful tentacles, which remind one of a lion's mane. They feed mostly on zooplankton, small fish, and moon jellies.

Range: Maine to Florida

Where am I? Lion's Manes are most often spotted in the late summer and autumn, when the currents bring them closer to shore. During low tide they will drift in the tidal creeks.

MANIAC ALERT

Avoid touching a Lion's Mane! Its sting will cause pain and redness and on occasion, severe burns. However, even though Sherlock Holmes in "The Adventure of the Lion's Mane" attributed a death to this creature, its sting is not considered fatal.

I don't sting! Like most comb jellies, the Common Northern Comb Jelly lacks the stinging cells that jellyfish have. They are often found washed up onshore and are no threat to beachgoers.

Common Northern Comb Jelly

Name: *Bolinopsis infundibulum*

Size: Up to 6 inches

Attributes: Though its body is transparent and gelatinous with a single body cavity, the Common Northern Comb Jelly is not a jellyfish and does not sting. Comb jellies also develop directly from fertilized eggs and do not have the developmental stages jelly fish do. The Common Northern Comb Jelly is the most common comb jelly in the shallow water of New England in the summer. Its clear body, though, makes it nearly invisible in water. It is extremely fragile and will almost always break when handled. Its body is oval-shaped and has two tentacles.

Range: Cape Cod to the Arctic

Where am I? They are found in the early summer. Their small clear bodies make them hard to spot, but they are most often seen when they are washed ashore.

Need your glasses?
The Striped Anemone is very common but easily overlooked because of its small size. However, once you spot one, it will usually have companions nearby.

Striped Anemone

Name: *Haliplanella luciae*

Size: Up to 3/4 inch in height

Attributes: Striped Anemone are minute animals identified by their green stalks with red to yellowish-white stripes. They feed on plankton, which they trap in their approximately 50 sticky tentacles. They appear stationary but, as with other anemones, move by gliding on their basal disc. Anemones reproduce sexually by releasing eggs and sperm, and reproduce asexually by leaving behind pieces of tissue when they move, that may grow into adult clones.

Range: A native to Japan, the Striped Anemone was first identified on the East Coast of North America, in New Haven, CT, in 1892. It now ranges from Maine to the Chesapeake Bay. It is also found in Europe.

Where am I? To find the Striped Anemone, bring the kids! It's helpful to have a child's eye close to the water to spot this tiny creature. Look for it attached to shells and rocks in tidal creeks.

Yummy? Though not typically eaten in the United States, the Common Periwinkle is a culinary treat in Europe. They are called "winkles" in Britain and Ireland, where they are traditionally sold cooked and salted, packaged in brown paper bags.

Common Periwinkle

Name: *Littorina littorea*

Size: Up to 1 1/4 inches in shell length

Attributes: The Common Periwinkle is found in huge numbers along the coast of the mid and north Atlantic. It was first discovered on the North American coast in 1840 and is believed to have been accidentally introduced from Europe. It has a multicolored gray and tan shell, that appears purple and dark gray when wet. It is a herbivore that feeds on algae it scrapes off rocks and grasses. At low tide it can exist outside water for hours at a time. To keep in moisture and to prevent drying out, it will enclose itself in its shell by creating a seal with its operculum.

Range: Nova Scotia to Maryland, though found worldwide and commonly in Europe

Where am I? At low tide, these gastropods are often found in groups in the mud, as shown in the photo at right, or attached to the blades of sea grasses, where they scrape off algae for food.

Cleans like a Hoover! The siphon of the Mud Snail (seen in the photo above, expanding over the surface of a girl's hand) acts like a perfect vacuum cleaner. The snail crawls over mud, consuming tiny pieces of animal remains.

Mud Snail

Name: *Ilyanassa obsoleta* or *Nassarius obsoletus*

Size: Up to 1 inch in shell length

Attributes: The Mud Snail is an abundant intertidal snail found in large herds. Its shell is dark brown to black in color. It crawls over the mud surface feeding on detritus and leaving a grooved trail behind it. The shape of its shell is less defined than most snails and appears worn down and eroded. Some think its name is derived from this less defined shape, as the Latin "obsoleta" means "worn out or decayed." It is preyed upon by ducks and the larger shorebirds.

Range: Native to the East Coast from the Gulf of Saint Lawrence to Florida. It was introduced around 1907 on the West Coast, where it is now found from San Francisco to British Columbia.

Where am I? To find a Mud Snail just look in the mud! This snail is abundant throughout most of the year. They love the mudflats and salt marshes at low tide and in sheltered waters.

No slipping here! If you try to pry loose a Common Slipper Shell from a communal stack, you will witness the strength of its suction-creating foot. Although it is able to move about on its foot, most will stay permanently attached to an object or stack.

Common Slipper Shell

Name: *Crepidula fornicata*

Size: Up to 1 1/2 inches in length of shell

Attributes: The Common Slipper Shell is a sea snail whose arched, rounded shell shows striations of gray, white, and pink. On the underside of the shell there is a plate across its width, about a third of its length, which resembles the toe cover of a slipper. Hence its name. The reproductive cycle of the Common Slipper Shell is unusual: When young, it is always male, but as it grows it may become female. It is often found in communal stacks, with the smaller males on top, larger females at the bottom, and hermaphrodites in between.

Range: Native to the East Coast, from Maine to Florida. It is also found in Europe.

Where am I? Look in the wrack line. As the tide pulls out, communal stacks are often found washed ashore.

Big Foot! The Moon Snail moves about on its incredibly large foot and burrows to find food. As it moves, it leaves a telltale trail showing where it has stopped.

Moon Snail

Also called Lobed Moon Snail and Shark Eye Snail

Name: *Neverita duplicata* (or *Polinices duplicata*)

Size: Up to 3 inches in diameter of shell

Attributes: The Moon Snail has a smooth, globe-shaped shell with a small spire. A line winds around the spire resembling a shark eye, giving it its other common name, the Shark Eye Snail. It is bluish-brown or purplish-gray in color. Its foot is extremely large relative to the size of its shell — so much so that it's hard to believe the snail can retract the foot. It is a carnivore and active predator. It feeds on other mollusks, using its radula to drill a neat and conspicuous hole through the prey's shell. Females lay eggs in distinct "sand collars," which they form out of mucus and sand grains.

Range: Cape Cod to Texas

Where am I? Look in tidal pools and creeks for trails like the one in the photo at left, or look for its victims. If you find a shell with a perfectly round, beveled hole, this snail is around.

An angler's favorite! The Clam Worm is popular as fish bait with New England fisherman and is harvested commercially for that use.

MANIAC ALERT

This bait worm has pincers and bites, so handle with care!

Clam Worm

Name: *Nereis virens*

Size: On Cape Cod, grows typically up to 8 inches in length. However, in other regions where the climate is colder, this species can grow as large as 3 feet!

Attributes: This is a spectacular though common species. Its color ranges from a greenish brown to a coppery red. Species of clam worms can be distinguished by their parapodia, the short limbs used for movement. On the Nereis virens, the parapodia are uniform from head to tail. The Clam Worm has a unique reproductive cycle. During lunar phases from late March to June, it undergoes heterogenesis. Its body changes — the parapodia enlarge so it can swim and it is capable of releasing eggs and sperm. They swim in breeding swarms at this time. Soon after they release their eggs or sperm, they die.

Range: Gulf of St. Lawrence to Delaware

Where am I? Look for small holes in the sand where they have burrowed. Dig gently as they are a bit fragile.

MANIAC ALERT

Be cautious picking up the Blood Worm — it's a feisty one. It will dart suddenly and with force. Its fangs are quite sharp. It can give a painful nip and even draw blood.

Blood Worm

Name: *Glycera* (multiple species)

Size: On Cape Cod, normally grows up to 6 inches in length and 1/4 inch thick, but one species can grow up to 2 feet long in certain regions.

Attributes: This worm has a semitransparent body that allows us to see the red body fluid that gives it color. On close examination, one can actually see the red body fluid pulsing and flowing to parts of its body. It has a round body with a tapered head and a large bulbous proboscis with four black fangs. Blood worms are sold for bait and are a favorite lure for flounder fishing. They are found in abundance on tidal flats, particularly mudflats. They burrow into the sand in a screwlike motion, aided by their large proboscis.

Range: Arctic to Argentina

Where am I? Blood Worms are abundant in mudflats. Though their burrow entrances are not evident in the shallow water, dig in a few areas of the sandy or muddy bottom, and soon you will find one.

Home decorating! To disguise its tube, the Plumed Worm will grasp nearby shells and rocks, and attach them to the sticky exterior of its tube.

MANIAC ALERT

Handle carefully. This worm can bite.

Plumed Worm

Also called Decorator Worm or Junk Worm

Name: *Diopatra cuprea*

Size: Up to 12 inches in length

Attributes: The Plumed Worm is identified by its tube which projects a few inches from the sandy bottom and is masked with shells, rocks, seaweed, and other debris. Hence its other common name, Decorator Worm. The tube opening is around 1/2 inch wide, and may extend 3 feet into the ground. While these camouflaged tubes are abundant on the mudflats, the worm itself is harder to spot. It will retreat quickly into its tube when disturbed. A carnivore, the worm will poke its head out of its tube when the tide is in to grasp passing prey. The worm has red plumelike gills on its head, and its body is strikingly iridescent, as shown in the photo at right.

Range: Cape Cod to Brazil

Where am I? Look for the decorated tubes in the tidal creeks. When water is moving over the tube, the worm might poke its head out to snatch passing food.

An abandoned home. The photo above shows a Bamboo Worm's tube. The worm lives upside down in its tube, with a small part of the tube extended above the sand. It uses its tail to close off the entrance to the tube.

Bamboo Worm

Name: *Clymenella* (multiple species)

Size: Up to 6 inches in length

Attributes: If you are lucky enough to find this worm outside its sand-encrusted tube, you will identify it immediately as the Bamboo Worm. Its segments look remarkably like bamboo cane. The ridges on the worm's segments are actually its parapodia, the extensions on worms used primarily for movement. The Bamboo Worm may be green or orange in color, depending on the species. Its head has no tentacles. It is able to regenerate should it lose segments. It lives in a tube it creates from cemented sand grains. To eat, it ingests sand and mud as it burrows, digesting food within and excreting the rest.

Range: Atlantic Coast from Maine to North Carolina, Pacific Coast from Canada to California

Where am I? The Bamboo Worm burrows in the sand of the mudflats but is not easy to find. It is more common to find its encrusted and abandoned tubes.

An architectural wonder! Abandoned Trumpet Worm tubes are often found on the beach. They are an amazing structure created from one single layer of sand cemented into a cone shape.

Trumpet Worm

Also called Cone Worm or Ice Cream Cone Worm

Name: *Pectinaria gouldii*

Size: Up to 2 inches in length

Attributes: The Trumpet Worm is a small worm that spends its life upside down in its tube. The cone-shaped tube remains buried except for the very end, which projects from the sand, and functions as a breathing tube. The Trumpet Worm's head is flat, and it has numerous golden bristles along its body. It and the other worms in this book are of the class Polychaeta, which means "many bristles." The Trumpet Worm will use its bristles to burrow into the sand headfirst and to sift through the sand for bits of plant and animal matter to eat. Once it digests its food, it excretes the waste back to the surface.

Range: Entire Atlantic Coast

Where am I? Dig in the sandy areas of the mudflats. It is more common to find their cones, but with persistence and luck you just might find the worm.

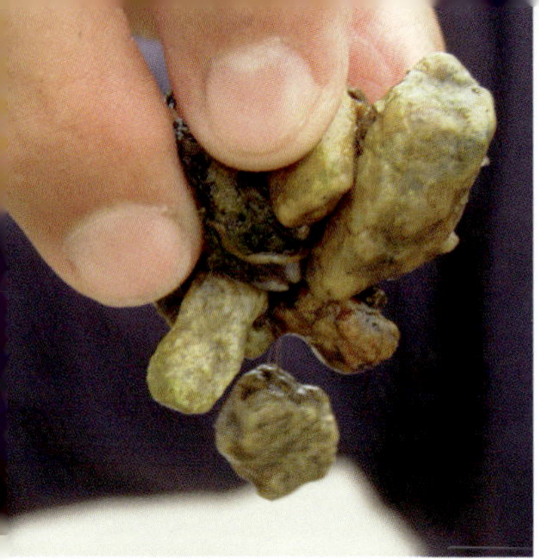

Sticky stuff! The foot of the Blue Mussel secretes tough threads, which are used to attach the mollusk to almost any solid object. In the photo above, a rock dangles from a few of these threads, while the mussel hides under other stones covering it.

Blue Mussel

Name: *Mytilus edulis*

Size: Up to 3 inches in length of shell

Attributes: An extremely common mollusk, the Blue Mussel can be found in extensive colonies in areas with rocks or pilings to which they can attach themselves. Their concave shells are smooth, with very fine concentric growth lines. The shell is bluish-black in color on the outside, and a light violet on the inside. They attach themselves to objects by tough threads created by the mussel's foot. When many threads have formed, they create what is referred to as a "beard." They are deliciously edible and are farmed and harvested for commercial use.

Range: Canada to South Carolina. Also in Europe and in other temperate and polar waters around the world.

Where am I? You will find these mollusks in groups, often attached to rocks. In fact, if you find a clump of rocks stuck together, there is a good chance there is a mussel inside holding the group together.

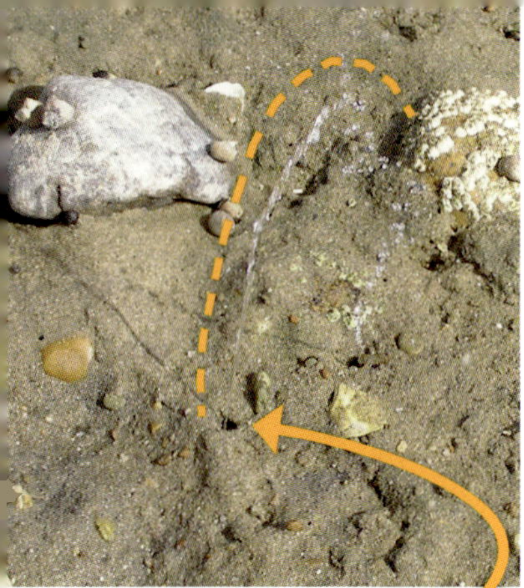

It squirted me! Like other clams, the Quahog's siphon takes in and expels water when it is buried in the sand. Periodically, it will shoot water out through the siphon. These tiny water fountains identify its location just inches below.

Quahog

Also called Hard-Shelled Clam

Name: *Mercenaria mercenaria*

Size: Up to 4 inches in width of shell

Attributes: South of New England the Quahog is aptly called the Hard-Shelled Clam. It is a culinary delight to shellfish lovers. Quahogs are sold commercially with different names depending on their size: as Little Necks when up to 1 1/2 inches; Cherrystones up to 2 inches; and Chowder Clams up to 3 inches or more. The shell is light tan in color with concentric lines. The inside of the Quahog is a lovely purplish color, which the Native Americans used for beads and currency, hence its species name "mercenaria." Its valve will shut tight when disturbed.

Range: Cape Cod to the Gulf of Mexico

Where am I? Stomp! When you locate burrow holes in the sand at low tide, stomp your foot to find a Quahog. The disturbance will make the clam shut tight and shoot water out its siphon, as in the photo at left.

Favorite food? Humans are not the only ones who love to eat the Soft-Shelled Clam. It is feasted upon by diving ducks, skates, and other bay creatures. Abandoned and broken shells can be found in abundance in the peat and along the shore.

Soft-Shelled Clam

Also called Longneck Clam and Steamer Clam

Name: *Mya arenaria*

Size: Up to 4 inches in width of shell

Attributes: The Soft-Shelled Clam is a favorite New England seafood, that is commonly fried or used in chowder. It is farmed commercially. Its distinct chalky-white shell is easily broken. It lives burrowed in the sand some 4 to 12 inches deep. While buried, its long neck acts as a siphon allowing water in and out of the clam. The siphon is encased by a fleshy tube that cannot be pulled back entirely into the shell and unlike other clams, the valves of its shell never completely close.

Range: Maine to South Carolina. Has been introduced to the Pacific Coast.

Where am I? Look in the mud and the peat along the sides of the tidal creeks. If you see water squirt from a half-inch burrow hole, the clam could be up to a foot below. While the Quahog is also a squirter, the Soft-Shelled Clam prefers muddy areas.

An Olympic diver? This clam has a fascinating dive into the sand. First, it extends its long foot into the sand, as this baby Razor Clam is doing in the photo above. Then, it widens the end of its foot to anchor itself as it pulls itself straight up, perpendicular to the sand. Amazingly fast, it will then dive straight down, disappearing into the sand, often in one quick motion.

Razor Clam

Also called Atlantic Jackknife Clam

Name: *Ensis directus*

Size: Up to 10 inches in length of shell

Attributes: The name Razor Clam is derived from the shape of this animal's thin shell, which resembles an old straight razor. The shell is long, six times longer than its width, and brown and black in color. It has a short siphon and lives just below the sand. When disturbed, it will squirt out water from its siphon. Its burrow hole is keyhole-shaped versus round. It moves in water by opening and closing its shell, and by stroking the water with its foot. This clam is edible and delicious. In addition to human predators, shorebirds and marine worms also prey upon the Razor Clam.

Range: Canada to South Carolina. Now also found in northwestern Europe, recorded there since 1978.

Where am I? Look in the sandy mud of the tidal creeks and pools. They are often found in colonies. But if you want to catch one, good luck. This clam is fast!

It's not a worm. The five distinct white muscle bands of the White Synapta identify this animal as an echinoderm. Below, its burrow hides under small mounds of sand.

White Synapta

Name: *Leptosynapta tenuis*

Size: Up to 6 inches in length

Attributes: The White Synapta may look like a worm but it is a Sea Cucumber and part of the Echinodermata phylum. It is a close relative to the sea star. It is identified by the 12 branched tentacles at its head and the five distinct white muscle bands visible along its transparent body. As with other echinoderms, the White Synapta can regenerate after losing pieces. A slow moving animal, it burrows in the sand and will project its 12 tentacles from the opening to pick up food. It feeds on detritus.

Range: Entire Atlantic Coast. Also found on the Pacific Coast and in Europe.

Where am I? White Synaptas are very common and can be found in abundance in tidal mud flats. To find their burrows, look for small mounds of fine sand, such as in the photo at left. Dig gently toward a burrow, starting a few inches away.

GLOSSARY

Arthropod: An invertebrate animal with jointed limbs that usually has a shell, and is a member of the phylum Arthropoda.

Asexual Reproduction: Reproduction by one animal that does not involve the union of male and female reproductive cells.

Basal Disc: A smooth muscular disc on the lower portion of an anemone used to attach the animal to an object.

Carapace: The large shell of a crustacean that covers its dorsal area.

Crustacean: An arthropod with a shell covering and a member of the class Crustacea, which includes crabs, lobsters, shrimp, and barnacles.

Detritus: Organic material broken down into fine particles by decomposition and the action of the ocean water.

Diurnal: Occurring daily or approximately every 24 hours.

Gastropod: A mollusk and member of the class Gastropoda, which includes snails and whelks.

Genus: A subdivision of a family or subfamily classification of animals normally containing more than one species.

Gonad: An organ in an animal that produces such reproductive cells as sperm or eggs.

Hermaphrodite: An animal that contains both male and female reproductive organs.

Heterogenesis: The process in which an animal takes on a different form in its life cycle and/or creates offspring different from itself.

Intertidal Zone: The area or region of an ocean that is exposed to the air at low tide and submerged by water at high tide.

Invertebrate: An animal that lacks a backbone.

Larva: The young or early form of an invertebrate animal.

Mollusk: An invertebrate animal with a soft body and hard shell covering of one or more pieces, and a member of the phylum Mollusca.

Molt: The shell of a crustacean that is shed when it grows a new shell.

Mudflat: A level coastal wetland formed where mud is deposited by tides and rivers and is exposed at low tide.

Nudibranch: A shell-less marine snail having external respiratory appendages.

Operculum: A hard plate that closes off and seals the opening of a gastropod's shell when the animal is retracted.

Parapodia: The paired appendages on many worms used for locomotion or to fan water.

Phylum: A classification of animals below kingdom and above class.

Plankton: A collective mass of floating or drifting organisms occurring in a body of water, primarily comprised of microscopic algae and protozoa.

Polychaete: A worm with parapodia that have many hairlike bristles and a member of the class Polychaeta.

Proboscis: An elongated appendage in the oral region of certain worms and gastropods that is used primarily for feeding.

Radula: A band of horny teeth found in most gastropods, used to break up food, scrape food, or bore holes in the shells of other mollusks.

Sandflat: A level coastal area with a sandy bottom that is exposed at low tide.

Wrack Line: The line of seaweed and other debris left by high tide, indicating the furthest reach of the tide that day, evident as the tide recedes.

Zooplankton: The collective mass of animal or animal-like organisms in plankton.

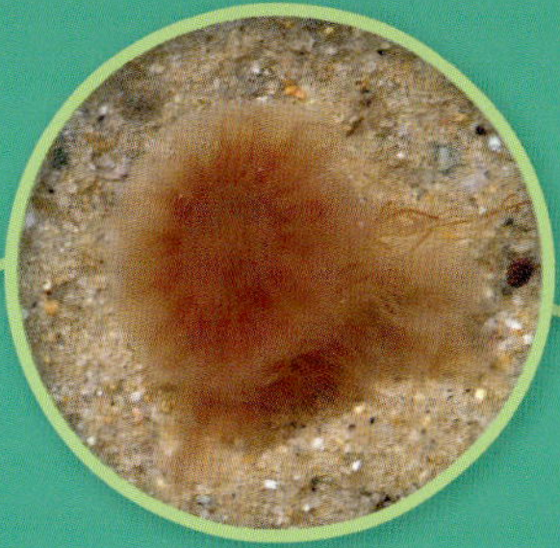

$1.49
866607
AO--AO-
294Y
No Exchange
Books
Animal & Nature
unique

Which of the animals shown above is
subject of a Sherlock Holmes murder mystery? Which animal
is the only source of a compound used to save human lives?

Inside ***Mudflat Mania!*** you'll find
the answers and other facts about
the fascinating creatures that
move the mud on Cape Cod.

ISBN: 978-0-615-23223-2
50895
9 780615 232232

$8.95 USA
ISBN 978-0-615-23223-2